# 欧式别墅新古典

# European Style Villa Design

深圳市艺力文化发展有限公司 编

大连理工大学出版社

**图书在版编目(CIP)数据**

欧式别墅新古典 / 深圳市艺力文化发展有限公司编
—大连：大连理工大学出版社, 2012.6
ISBN 978-7-5611-6908-7

Ⅰ.①欧… Ⅱ.①深… Ⅲ.①别墅—建筑设计—图集
Ⅳ.①TU241.1-64

中国版本图书馆CIP数据核字（2012）第081088号

---

出版发行：大连理工大学出版社
（地址：大连市软件园路80号 邮编：116023）
印 刷：利丰雅高印刷（深圳）有限公司
幅面尺寸：245mm × 290mm
印 张：21.5
插 页：4
出版时间：2012年6月第 1 版
印刷时间：2012年6月第 1 次印刷
责任编辑：裘美倩
责任校对：王秀媛
特约编辑：丁小娟
装帧设计：王 薇

---

ISBN 978-7-5611-6908-7
定 价：320.00元

电 话：0411-84708842
传 真：0411-84701466
邮 购：0411-84703636
E-mail: designbook@yahoo.cn
URL: http://www.dutp.cn

如有质量问题请联系出版中心：（0411）84709043 84709246

# European Style Villa Design

# Contents

## Ken Tate

## Costermans Villa Projecten Belgium

## James Glover

## Joseph N. Campitelli Architect Inc.

## London Sotheby's International Realty

## Drew Mandile | Brooke Knapp

# Hay

Design Agency: Ken Tate

Set atop a small mountain in the Kentucky horse-breeding country, this 12,000 s.f. compound is an unalloyed expression of Federal architecture. Constructed of hand-made brick, the house's façade features a central portico with cut-limestone columns and a pediment with an oval oculus. Similar porticos cap identical wings flanking the façade to create perfect balance and an air of tranquil repose. Clearly visible from every side of its hilltop perch, this country retreat offers an equally serene rear façade, where only a wooden sunroom interrupts the symmetry. Within, restrained Regency style details perpetuate the dwelling's air of calm elegance. Reiterating the Classical language of the estate in its barn, Tate drew inspiration from the portico of Inigo Jones' Covent Church in London for simple Tuscan columns that support a pediment with a round oculus.

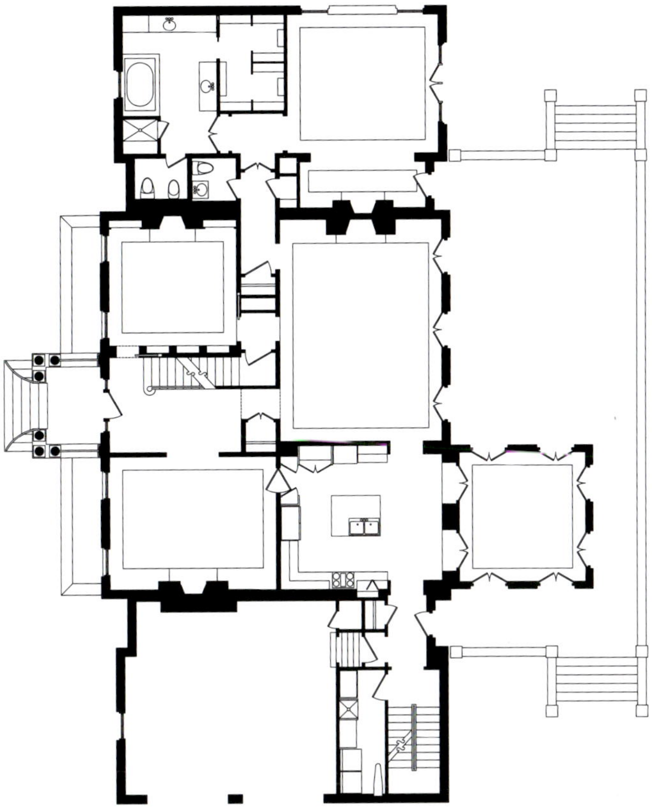

BARN
The Old Barn Book
FARM
MONET
THE IMPRESSIONISTS

# Anderson

*Design Agency: Ken Tate*

The New World meets the Old in this 7,000 s.f. residence rising above the sandy beaches of Gulfport, Mississippi. Defined by graceful symmetry and an inviting sense of ease, the house is a modern marriage of Palladian and French West Indian architecture. While the form—a central block surrounded by four pavilions—dates from Palladio's time, the slate-covered hipped roof is a hallmark of French Colonial plantations. Throughout the dwelling, louvered shutters filter sunlight and sea breezes, invoking the comfort and romance of these island dwellings. Within, refined French continental shapes including segmental arches, carved limestone mantels, and cove moldings, recall European elegance. But the absence of highly decorative detail and choice of a monochromatic palette gives the house a unifying, and contemporary aspect.

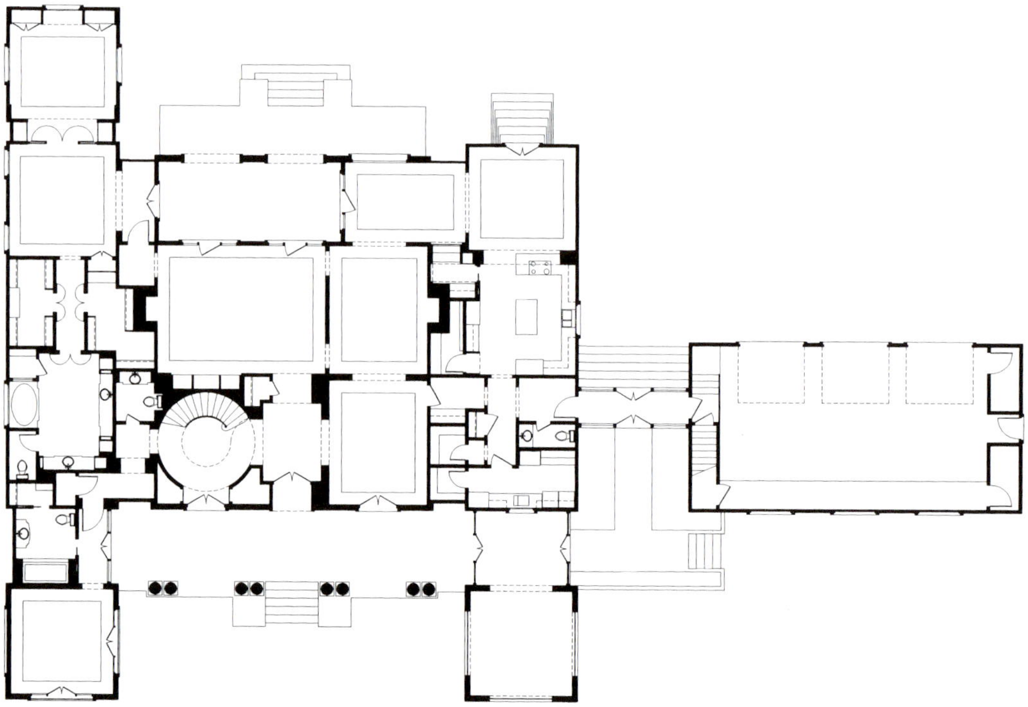

# Alexander

*Design Agency: Ken Tate*

Capped with a low, hipped roof and surrounded by deep porches, this 12,000 s.f. residence in New Orleans resembles a Creole plantation in the French West Indies. A Federal fanlight over the front door and Spanish Colonial-style ceiling beams and arches in the living room serve as reminders of the Anglo-American and Spanish influences that merged with French styles to create New Orleans' unique architectural character. Wings resembling outbuildings are attached to the central block of the house by low infill walls—a technique used in Colonial times to join pre-existing structures. Details like these reveal the architect's fascination with old buildings that grow and change over time and his interest in creating new ones that capture the scale and atmosphere of such dwellings.

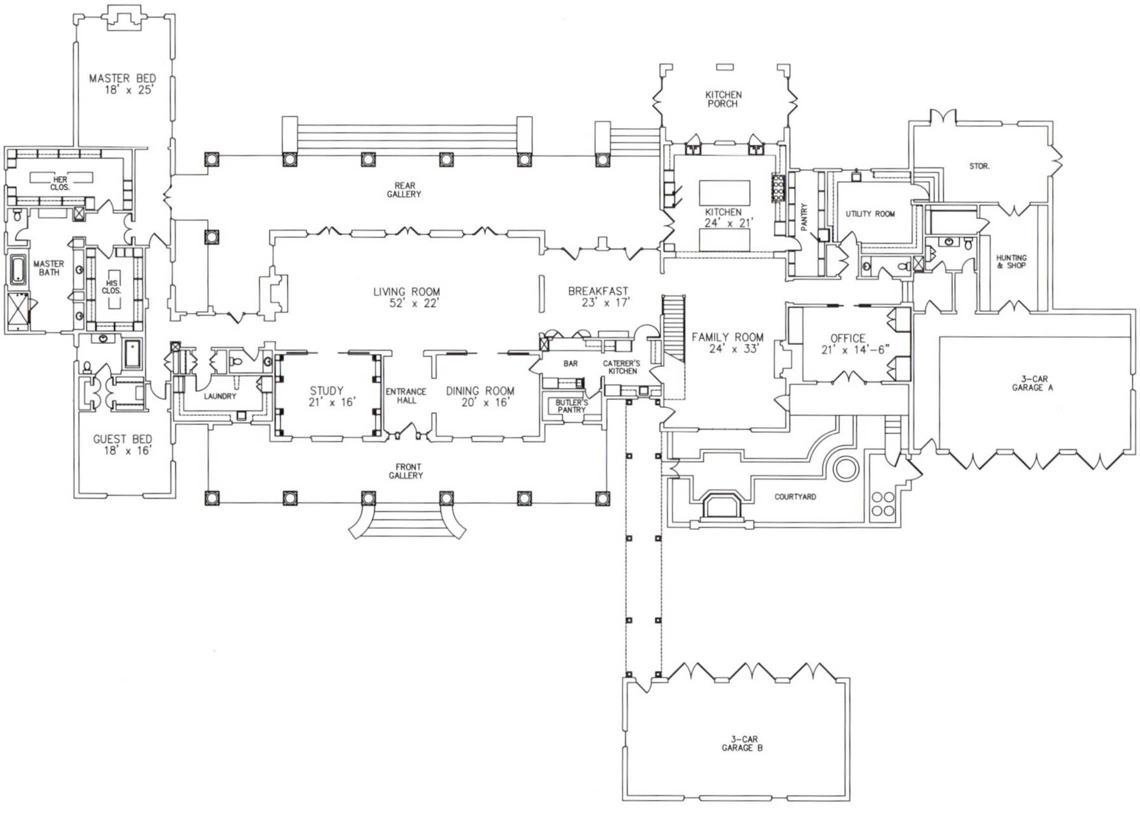


FIRST FLOOR PLAN

# Randolph

*Design Agency: Ken Tate*

Tate's interest in marrying varied Classical styles finds full expression in this 9,000 s.f. house in Oxford, Mississippi. Inspired by a Federal house in Natchez, but executed in the Greek Revival style favored in Oxford, this residence offers a restrained, slightly feminine rendition of this exuberant Southern style. While delicate Ionic capitals lend the front porch a delicacy more in keeping with Federal design, heavy stucco-over-brick columns with Doric capitals create pure Greek Revival façades for two rear outbuildings. The interior expresses Tate's interest in creating fictional "back-stories" to imbue new houses with the character of old dwellings. Implying that the house was once a one-room deep Federal dwelling with a rear porch, the study and dining room open through tall double-hung windows into a long living room resembling an enclosed porch.

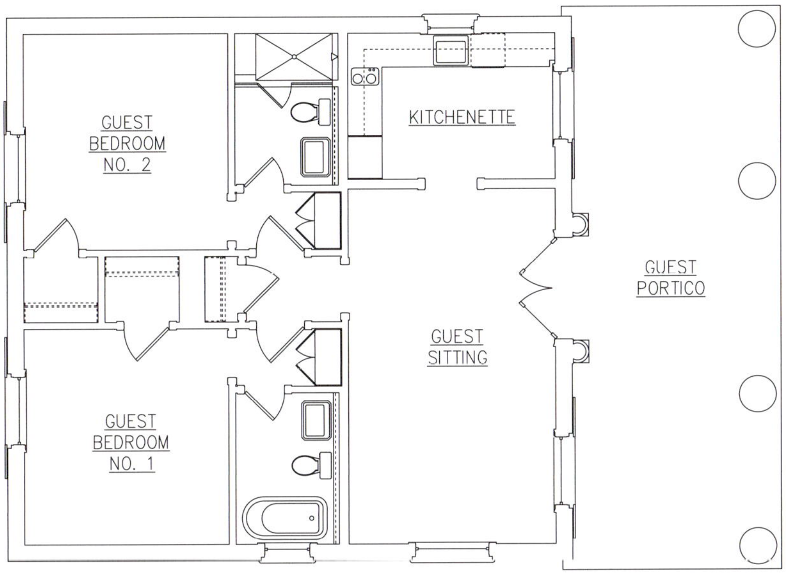
GUEST BEDROOM NO. 2
KITCHENETTE
GUEST PORTICO
GUEST SITTING
GUEST BEDROOM NO. 1

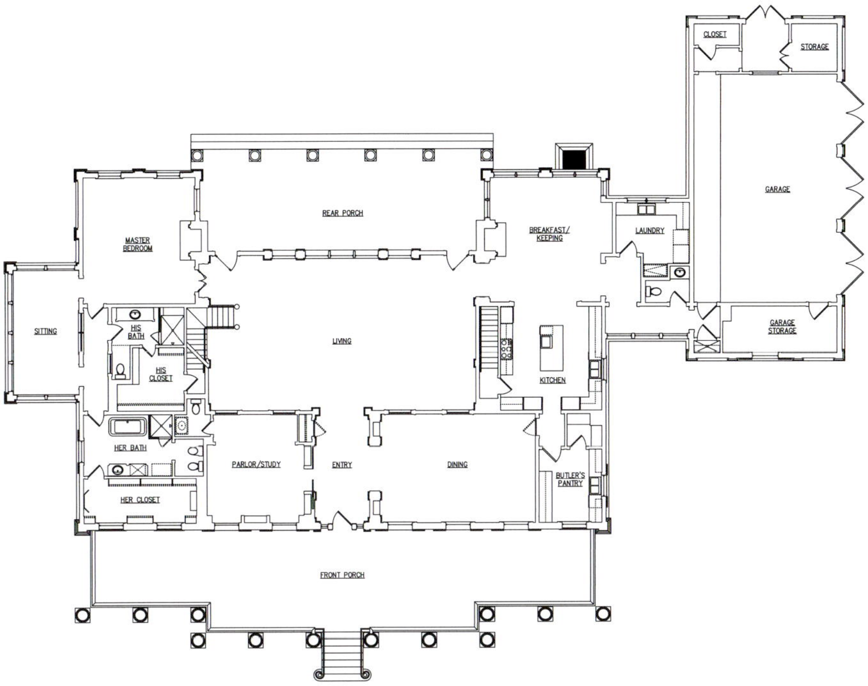
CLOSET
STORAGE
GARAGE
REAR PORCH
BREAKFAST/ KEEPING
LAUNDRY
MASTER BEDROOM
SITTING
HIS BATH
HIS CLOSET
LIVING
KITCHEN
GARAGE STORAGE
HER BATH
PARLOR/STUDY
ENTRY
DINING
BUTLER'S PANTRY
HER CLOSET
FRONT PORCH

# Smith

*Design Agency: Ken Tate*

To create this 12,000 s.f. compound in Houston, Texas, Tate drew upon his intuitive grasp of Creole architecture. Designing several wings and outbuildings, he employed styles and materials that reflect varied moments in the evolution of Creole design. A children's house shaped like a pigeonnaire and a freestanding study with a steep hipped roof evoke French West Indian plantations, while the encircling wall, Spanish arcades, and limed brick wing suggest the Spanish influence. With Creole details, including several French doors opening from the porch into first-floor rooms, combined with the Federal-style central entrance and Greek Revival interior moldings, the main house reveals 19th-century Creoles' interest in American styles. Suggestive of a single family's changing tastes, the house is imbued with an aura of history and romance.

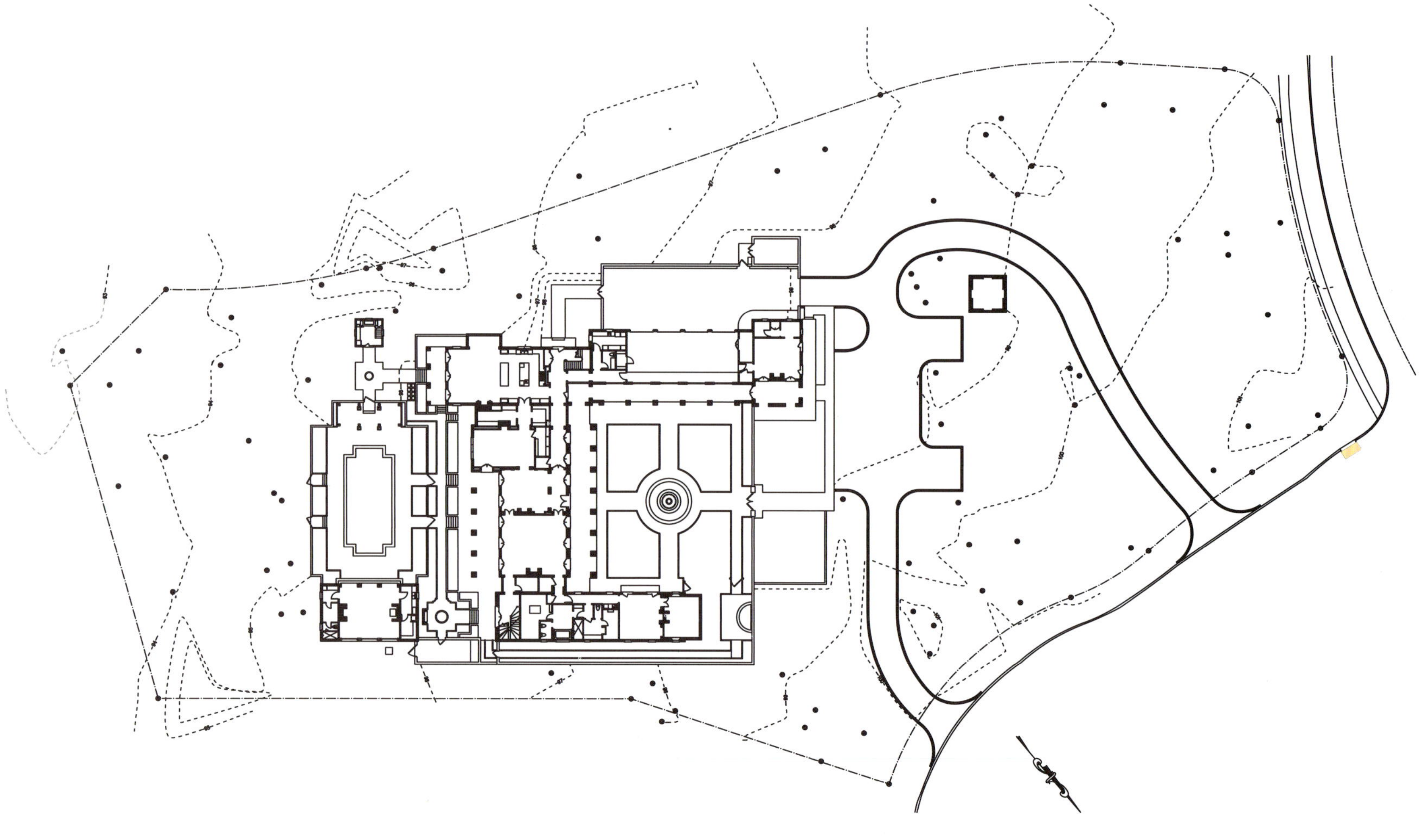

# Aycock

Design Agency: Ken Tate

LIFE

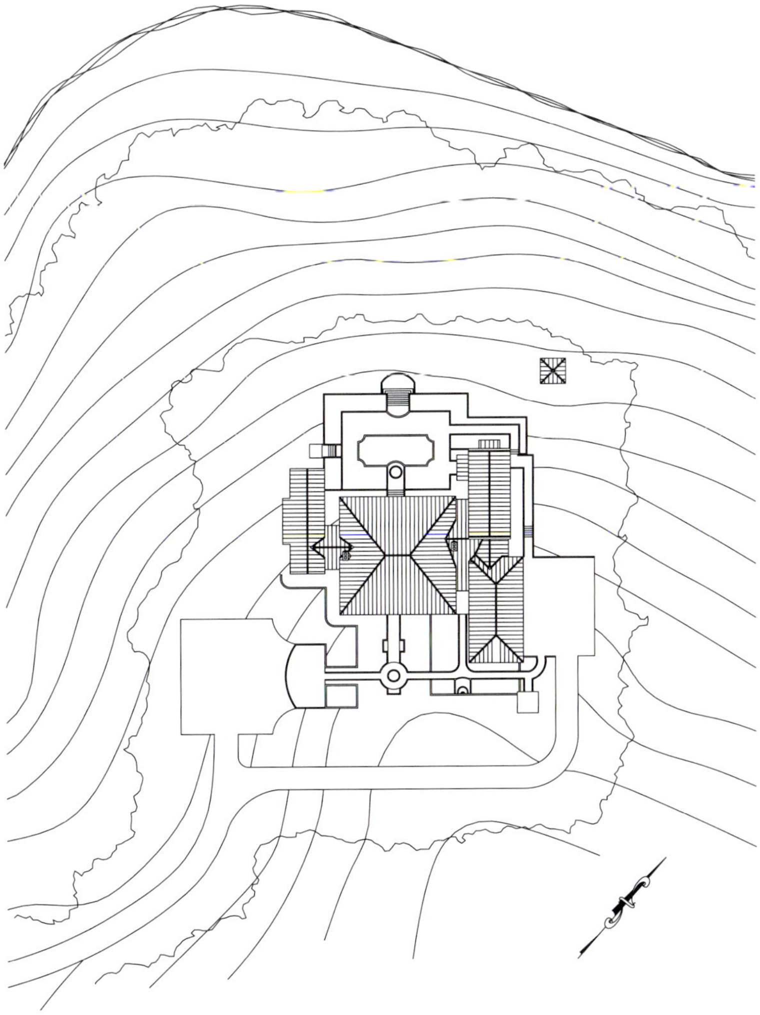

# EMERY

*Design Agency: Ken Tate*

This 10,000 s.f. residence in the historic Belle Meade neighborhood of Nashville, Tennessee invokes the concept of vernacular progression—the manner in which houses grow and evolve over time. Disparate parts of the dwelling resemble a vernacular Tennessee fieldstone barn, a Georgian house, a Federal wing, and a Colonial Revival addition. Thanks to Tate's attention to period-appropriate materials and craftsmanship, these attached structures appear to have been built over two centuries. This approach made it possible to design a large house accommodating the owners' modern needs, while also complementing the style and scale of the surrounding Colonial Revival dwellings. Although the house is significantly larger than those nearby, the manner in which Tate divided the mass into smaller parts diminishes its appearance of size.

AMERICAN GLASS
1760-1930
AMERICAN GLASS
1760-1930

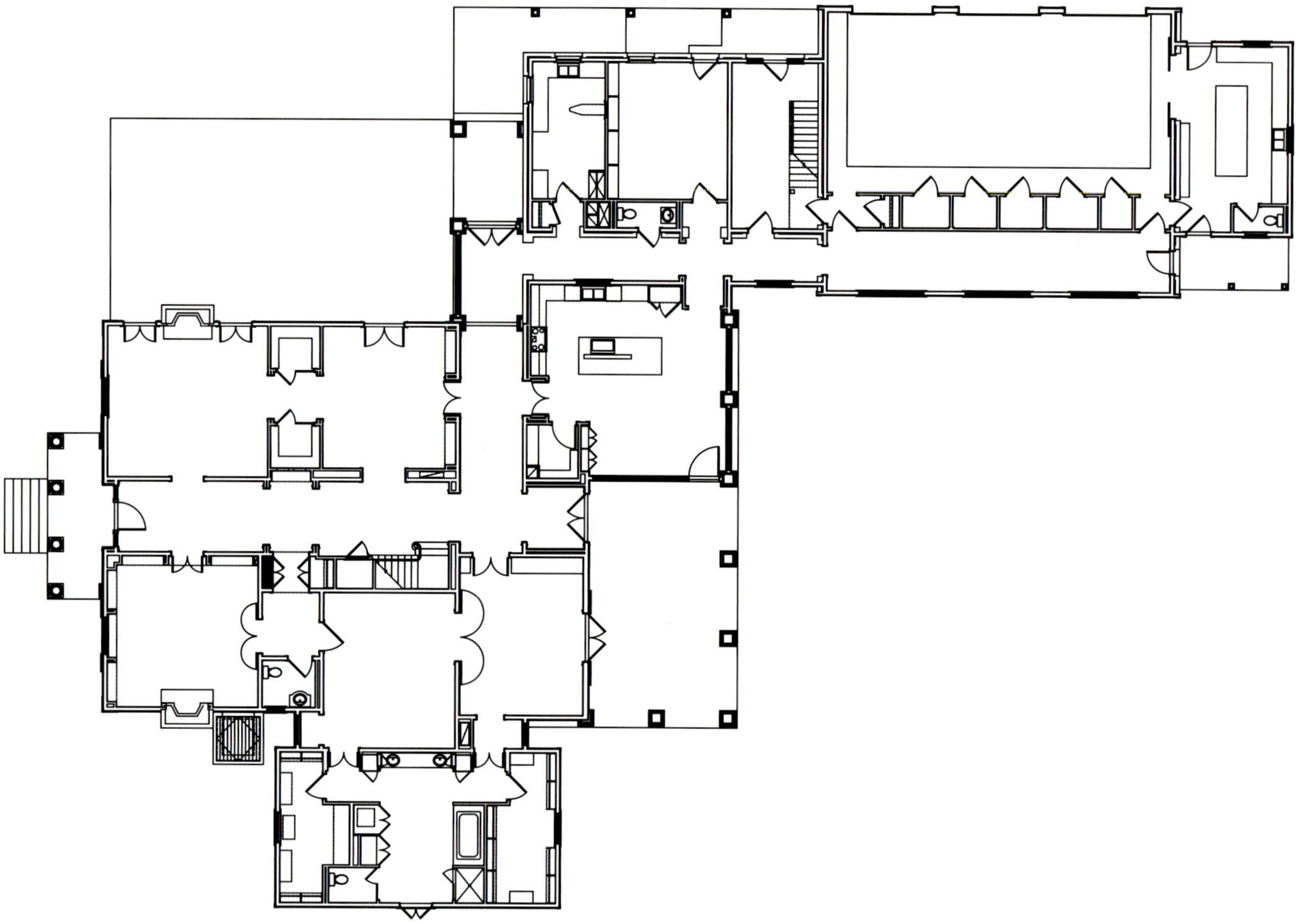

NATIONAL GEOGRAPHIC ATLAS OF THE WORLD

# Imes

*Design Agency: Ken Tate*

In this 11,000 s.f. house outside Columbus, Mississippi, Tate inventively resolved his client's wishes for a French provincial style residence and a Neoclassical interior. Rather than create a seamless marriage, the architect chose a collage-like layering of high-style and low-style elements. The resulting house appears to be a primitive Renaissance-era dwelling remodeled in the neoclassical style. Upon approach, the rough-hewn façade and hip roof resemble a simple, Norman farmhouse. But the heavy oak door opens into an elegant foyer with trompe-l'oeil decoration and a plaster dome. A similar juxtaposition occurs in the living room, where rough-hewn ceiling beams appear to predate refined lime-coated paneling. This layering of the rustic and refined finds full expression on the rear façade, where an elegant turned-stone balustrade surmounts rubble stone stairs.

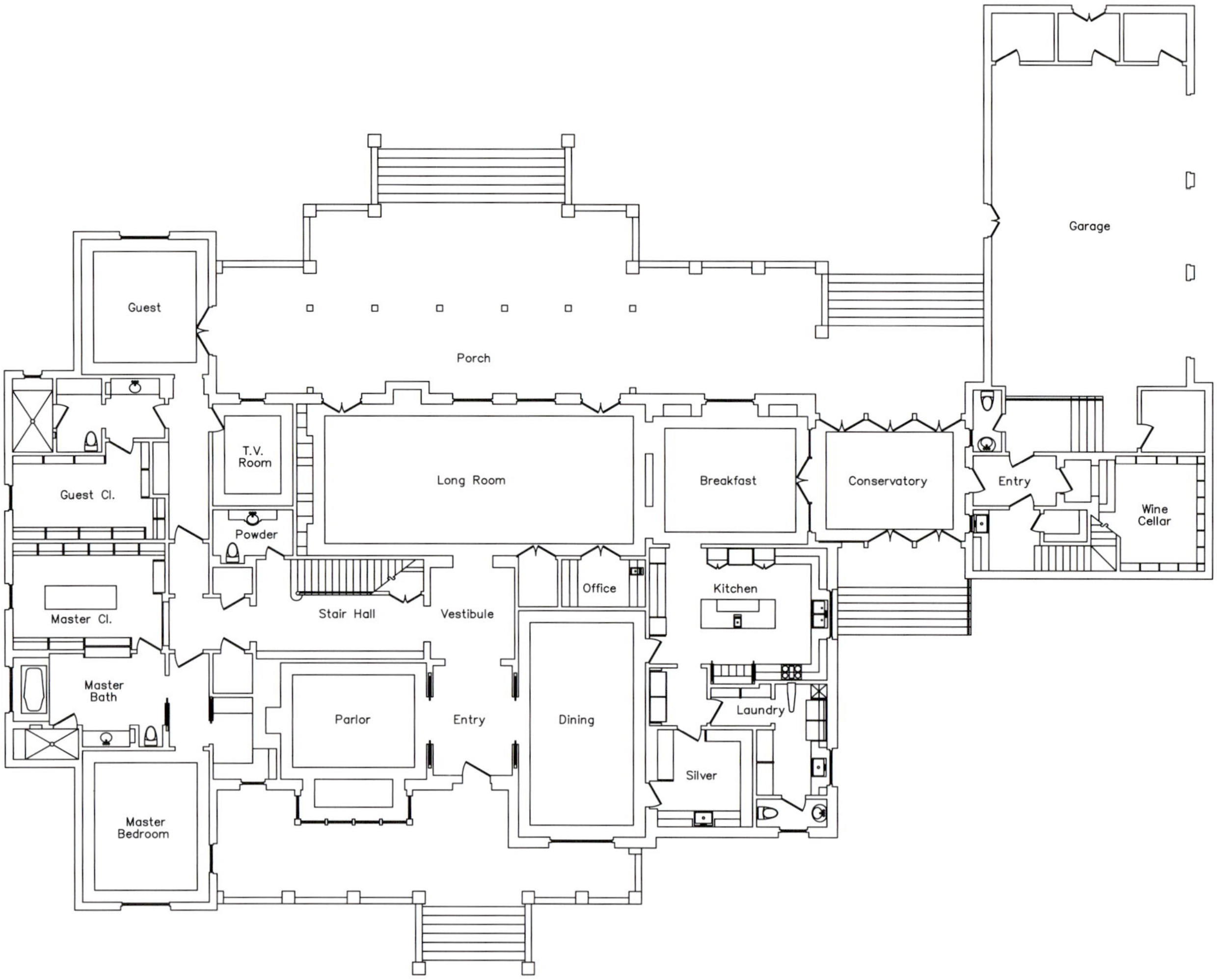
Garage
Guest
Porch
T.V. Room
Long Room
Breakfast
Conservatory
Entry
Wine Cellar
Guest Cl.
Powder
Master Cl.
Stair Hall
Vestibule
Office
Kitchen
Master Bath
Parlor
Entry
Dining
Laundry
Silver
Master Bedroom

# Norton

*Design Agency: Ken Tate*

Glimpsed through stucco-and-stone entrance pillars, this 4,000 s.f. house resembles the modestly elegant maisons that 18th-century French nobility built as country getaways in the suburbs of Paris. The façade's pristine natural white stucco, French doors with dark shutters, and steep hipped roof offer a distinctly northern European impression of discreet and measured charm. In contrast, the rear façade invokes the romantic spirit of a hillside villa in Provence or the Riviera. Built upon a slope that descends dramatically to a lake, the house has a multi-level rear façade with wide windows and glazed doors, several balconies, and a balustrade terrace that afford expansive views of the landscape. A diminutive pavilion with Tuscan columns and a cedar-shingled roof adds a touch of whimsy at the edge of the lake.

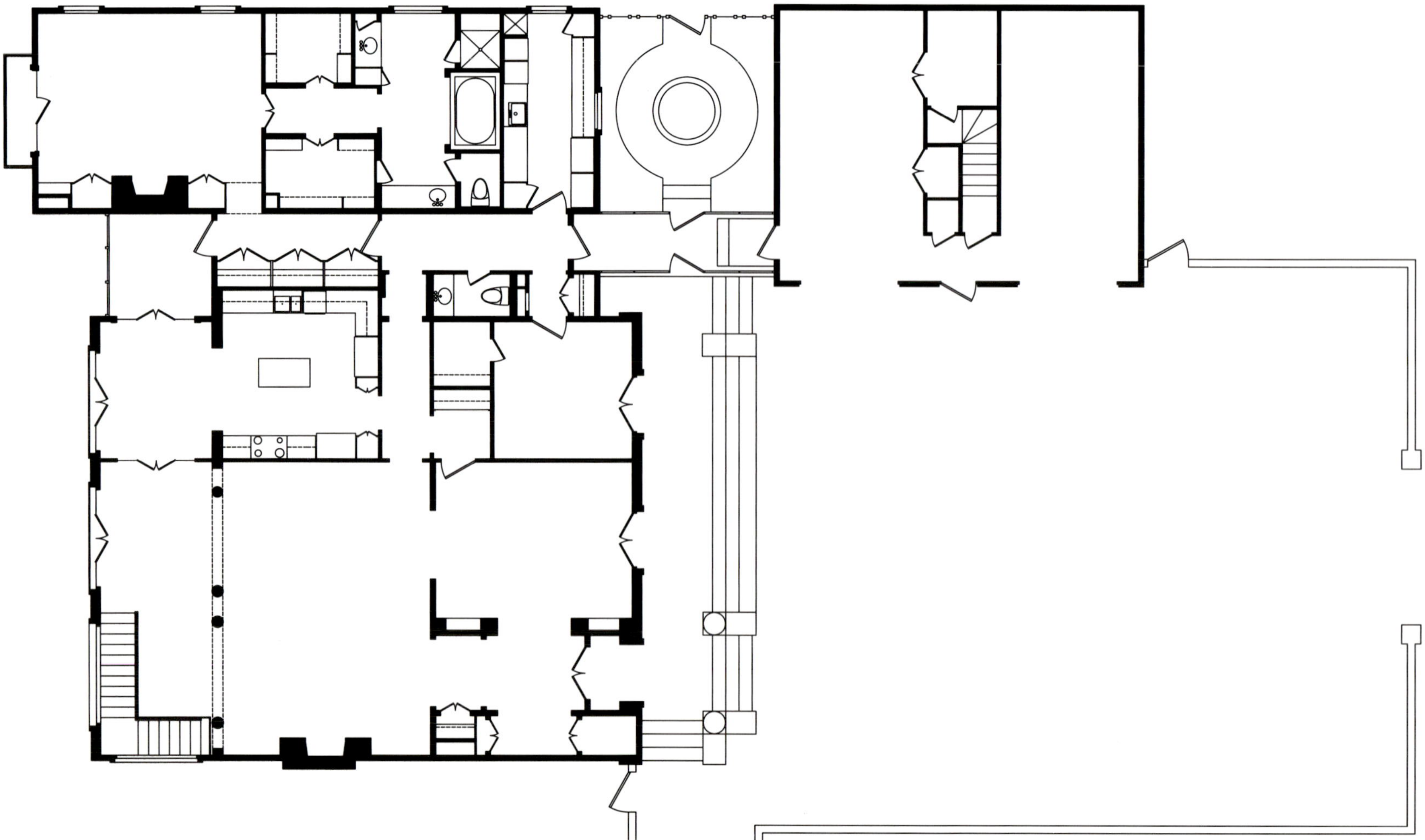

# Stanford

*Design Agency: Ken Tate*

In this 7,000 s.f. residence in Jackson, Mississippi, Tate employed a variety of styles to invoke a range of emotions. Approached by a courtyard, the stucco and stone façade—complete with turret—resembles an old Norman dwelling. While this style lends the street-side façade an introverted, protective attitude, a Mediterranean-style loggia overlooking a lake creates an extraverted mood on the rear. Within, rooms include an antique stone-and-cabochon paved entrance hall, a dining room with lime-coated paneling, and a massive living room with large French doors opening to the landscape. The dwelling invites residents to take a daily journey through an iron gate into the reassuring enclosure of the courtyard, into the welcoming embrace of the entrance hall, and finally toward the uplifting aspect of the lake.

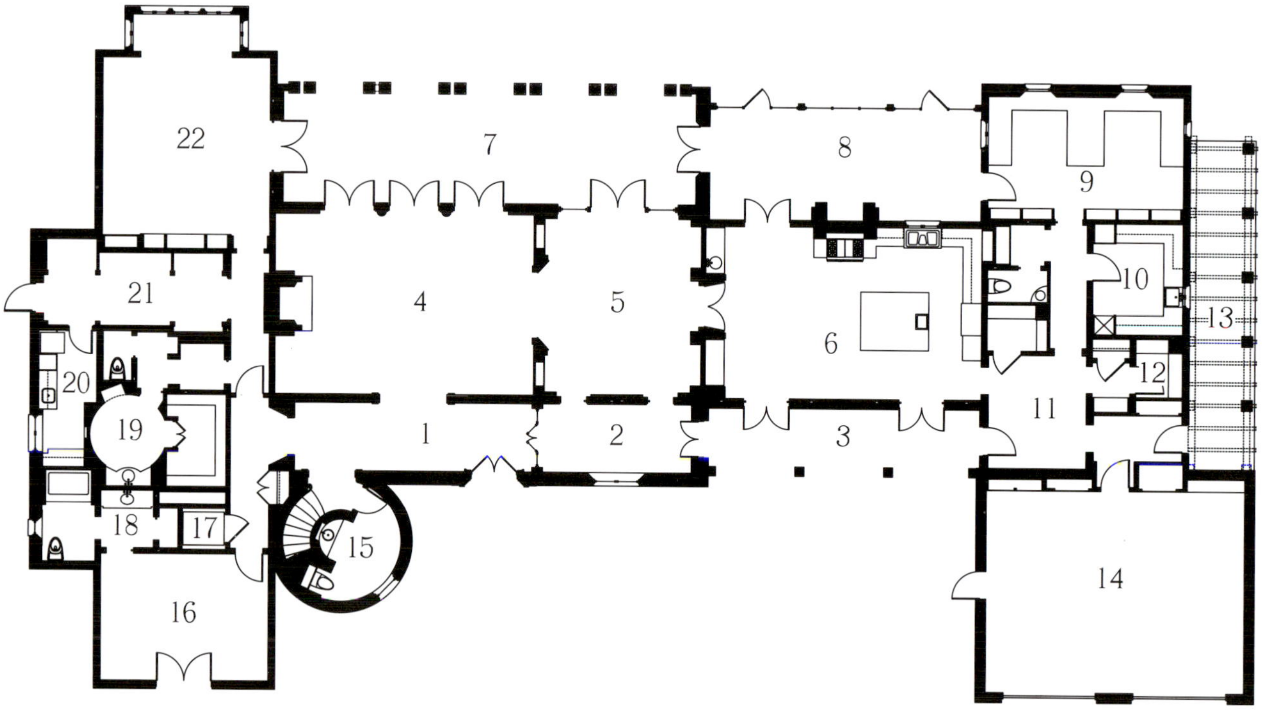

FIRST FLOOR PLAN

1 ENTRANCE HALL
2 GALLERY
3 PORCH
4 LIVING
5 DINING
6 KITCHEN BREAKFAST
7 LOGGIA
8 SCREENED PORCH
9 OFFICE
10 LAUNDRY
11 SIDE ENTRY
12 WINE CELLAR
13 PERGOLA
14 GARAGE
15 POWDER
16 LIBRARY
17 ELEVATOR
18 BATH
19 BATH
20 LAUNDRY
21 GALLERY
22 MEDIA ROOM

# *Young*

*Design Agency: Ken Tate*

In the tradition of sixteenth-century Italian architect Andrea Palladio, this 5,000 s.f. residence in Baton Rouge, Louisiana marries the grandeur of Roman temple forms with the comfortable intimacy of a lakeside villa. The portico's tall columns and noble geometry create an imposing initial impression. But the protective embrace of rustic style wings flanking the forecourt, combined with the ochre-colored stucco and Roman-style red tile roof, creates an invitingly human-scale environment. On the rear, a second-storey balcony and deep loggia face views of a natural oak allée leading to a bayou. While the overall form and exterior design of the house draws directly from Palladio's country villas, the design of the grand salon and dining room within recalls the more highly embellished decoration of 17th-century urban Roman dwellings.

VERSAILLES

Master Bedrm.
Loggia
Kitchen
Cl.
Butler's Pant.
Brkft.
Her Bath
Dress.
Salon
Entry Hall
Dining Rm.
Pant.
Laund.
His Bath
Stor.
Cl.
Portico
Library

# Cotswold Cottage

*Design Agency: Costermans Villa Projecten Belgium*

The characteristic cottages, numerous present in the English Cotswolds, were the inspiration for Costermans Villa Projects to design this spacious home in Antwerp. At the outside, all typical architectural elements of the cottage are present. The Interior got a more contemporary interpretation: timeless, sober and light.

The interior is much lighter than the traditional cottages, with warm materials. Old materials as terracotta tiles in different sizes and laying patterns, Burgundian tiles, a sandstone fireplace, etc. are used throughout the house.

# English Manor House 1

Design Agency: Costermans Villa Projecten Belgium

Although the house is new, it seems that people have been living here for years. The rustic, rural style of the interior and the numerous authentic materials make the home into a charming haven where time stands still. The home becomes an ideal place to relax after a busy workday and to enjoy the warmth of family togetherness.

One of the most principal characteristics of this house designed by Costermans Villa Projects is the French, English style. Everything is nicely in balance. The base of the interior is formed by many anthieke building materials such as an empathic parquet of old French oak, blue limestone in combination with Moroccan Zelliges, a checkerboard pattern of sandstone and shale, old fireplace, etc.

# English Manor House 2

*Design Agency: Costermans Villa Projecten Belgium*

Within a setting of lovely park trees, Costermans Villa Projects realised a timeless English Manor. The architectural design and use of materials have strong English roots. Old bricks (baked in open ovens), oak windows, recuperated roof tiles and elegant wrought ironwork melt into a harmonious and timeless whole. The garden architecture was approached in a playful way in order to respect its parklike character as much as possible. When making a stroll through the garden, you will meet a renovated pond surrounded by cobbles from the Meuse River: a closed meander of the nearby brook.

The character of the interior can be described as follows: a harmonious whole of old materials, wrought ironwork and warm colours.

# French Manor House 1

*Design Agency: Costermans Villa Projecten Belgium*

This classic French manor in the leafy outskirts of Antwerp, Belgium was constructed by Costermans Villa Projects.

The company devoted the utmost attention to the selection of materials, installing unusual antique tiles, durable old parquet floors, exceptional fireplaces and other special details. The look is timeless, yet contemporary: a sober and sophesticated living environment.

6 MURA

LENNON

TOM FORD
JAZZ LIFE

# French Manor House 2

*Design Agency: Costermans Villa Projecten Belgium*

This is a new villa with French History.

France remains the inexhaustible inspiration for rural homes. This timeless villa by Costermans Villa Projects is a perfect example.

The house not only exudes an atmosphere of warmth, but the majority of the rustic materials are of French origin and constructed the home in a Belgian fashion.

# Belardi Residence

*Design Agency: James Glover*

Atop a scenic Southern California coastal bluff, the Belardi Residence commands stunning Pacific Ocean views punctuated in the evenings by sea breezes and spectacular sunsets. On a triangular parcel, the plan is meticulously formed to take advantage of the view and respond to the microclimate of an adjacent coastal canyon. The home's open planning allows space to be utilized in discreet units defined by the furnishings or opened up to provide space for large scale entertaining. The home's contours embrace a lush entrance courtyard serving as a ventilation nexus for the surrounding room spaces, allowing natural ventilation over a wide range of weather conditions and providing a private, sheltered space for outdoor living.

Every detail of the home has been carefully considered to enhance it's Italianate charm, such as the entry door with it's distressed wood frame, antique glazing and hand crafted wrought iron detailing, providing a glimpse to the ocean vista just beyond. While utilizing the historic language of classical columns, elliptical arches, massive walls, stone trim, and rustic tile roofing, he still maintains the open space and airiness of a modern floor plan. Achieving this balance of the ancient and the modern is one of the central strengths of his style; as if the building could not have been conceived any other way.

# Colombatto Residence

*Design Agency: James Glover*

How must Ptolemy have felt surveying his domain from the terraces of the Library of Alexandria? We can only speculate, but it must have been close to the experience of living at the Colombatto Estate in San Clemente, California; directly adjacent to the "Western White House", home of President Richard M. Nixon. With 150 meters of ocean bluff frontage and 255 degrees of horizon view, this parcel is arguably one of the finest pieces of residential property on Earth.

The home created by Mr. Glover, in the Majorcan style, is not so much built "on" the site as it is "wrought in place" from materials as ancient and elemental as civilization itself. To achieve this level of finish and historic fidelity requires studied consideration of every detail to maintain the expected comfort level of a modern dwelling while still possessing the soul of it's monumental predecessors. The sprawling plan is focused around a massive entry turret which opens out to large spaces befitting the clients' requirements for entertaining large groups for charitable functions, while also providing private, intimate spaces for family living, arts, music or meditation. With expansive views of nature in it's fullest grandeur, from virtually every room, it's hard not to stop and contemplate things beyond the one's own place and time in this spectacular home.

# EADS

*Design Agency: James Glover*

Spanish Colonial Revival, a California trademark, is best done by a California native son. A lifelong resident of San Clemente, Mr. Glover has been immersed in historic Spanish Colonial Revival homes his entire life. San Clemente boasts over 100 of these unique buildings from the 1920's; half of them within walking distance of Mr. Glover's home.

On a bluff, overlooking the Pacific, the sun sets over Santa Catalina Island on the western horizon. True to its genre, the home's plan is a Villa, with all rooms arranged around a central courtyard that serves as an entry to the dwelling, a sheltered outdoor family gathering space and an area to entertain large groups. The courtyard is in reality an extension of the indoor space, complete with its own al fresco dining area, hearth and outdoor kitchen.

The finest detailing can be seen throughout the home, as demonstrated in the authentic waxed Spanish terra cotta flooring, and locally crafted hand-painted tiles surrounding the hearths of the Living and Dining rooms. Outside, the undulating white plastered walls, handmade terra cotta pan-and-barrel clay tile roofing, rusticated wood rafter tails and wood trellis work recall the sumptuous comforts and relaxed sensibility of the Jazz Age.

# Healy Residence

*Design Agency: James Glover*

The Healy Residence pays homage to the "Monterey" vernacular of the Spanish Colonial Revival style, in response to the long, wedge shaped parcel with views of the Pacific across pristine coastal chaparral to the sea. The garage wing occupies the front portion of the site and provides a spatial buffer between the residence and the street. At the center of the site, the entry opens to a central hall forming the middle leg of the "U" shaped plan which embraces a central courtyard open on the west side to a panoramic view. The plan provides privacy, frames the views from all rooms and gives exterior wall space to provide ample, natural ventilation for all interior spaces.

Constructed of cast-in-place insulated concrete, the building substantial in mass, very quiet, is extremely energy efficient and virtually indestructible by the forces of man or nature. Employing traditional Spanish Colonial materials and detailing, the home exhibits smooth plaster finishes, oil-rubbed millwork, handmade roof tiles, natural stone, wrought iron balconies, concrete corbelling, and a daring cantilevered spiral stair. All casework was crafted from a single, ancient walnut tree milled and shipped from New Zealand by the client.

# Kahn Residence

Design Agency: James Glover

Situated on a small interior parcel, in a dense neighborhood with restrictive height limits, the Kahn Residence is an exploration of the rustic Italianate style in a compact building with limited frontage. On entering the home, one discovers a masterful composition of mass and form, curling around a central, two-storey café courtyard, filling the house with fresh air and light perfectly filtered by reflection from natural stone walls. An outdoor hearth stands ready for duty in the southeast corner, its cylindrical form rising to meet a flying buttress engaged to the home's east wing. An almost hidden, yet ample, stair quietly passes below the buttress providing passage to an outdoor kitchen and dining area at the main level, adjacent to the kitchen.

The west wing faces a panoramic view of the Pacific. The site's slope allows the lower level master and secondary bedroom suites to open fully to the view and a landscaped patio. At the main level, the kitchen, dining and living rooms are defined by a Great Room that is articulated into independent spaces by antique heavy timber beams and meticulously crafted walnut millwork. The Great Room opens onto a wide, decadently brick-paved terrace via a full-width sliding door, making the west wall open completely to the coast; true indoor-outdoor living.

# Manian Residence

Design Agency: James Glover

A Mediterranean landscape and mountain views give this Coto de Caza, California property the feel of the Italian country side, a perfect canvas for an authentic, Italianate, country home.

The hand-made quality of this home is immediately apparent as one enters through the large, meticulously carved walnut entry doors, into a grand foyer with a domed ceiling of reclaimed Italian terra cotta mosaic. Three arched passages lead to the main rooms of the house, including a music room directly ahead of the foyer, with an arched window invoking disciplined classical proportions. The main circulation path is through a groined passage that leads to the family room and kitchen, both sheltered below an expansive, brick, barrel vaulted ceiling. All of the main rooms of the home open to an expansive rear patio area landscaped in native plant materials.

Authentic hand craftsmanship shows up in all aspects of the work, from the wall stone which was cut on site, to the delicate ironmongery, to the intricate woodcarvings created expressly for this home from Mr. Glover's drawings and specifications. The building elements are massive, importing the spirit of Old World stone construction. The Manian residence recreates a rustic, Italian villa living experience, and only one hour away from Los Angeles.

22901

# Colombatto Cove House

*Design Agency: James Glover*

Situated on a narrow, yet deep, interior lot in a moderate density private neighborhood, this Spanish Colonial Revival style home utilizes a courtyard plan to provide private outdoor living space made available to the surrounding ground level rooms through richly finished French doors. The courtyard contains a large fountain, a kitchen garden and an outdoor dining and entertaining area, all arranged in formal symmetries with the surrounding building elements. The exterior finishes are in keeping with the vernacular with smooth plaster and distressed wood, along with some special touches including custom shaped rafter tails at the exposed eaves and delicately wrought iron work completed to Mr. Glover's specifications.

Inside, the millwork is carefully fit and finished to standards rarely seen in modern times. All doors and windows are genuine wood framed assemblies with modern performance, but finished with classic early 20th century antique stain finishes. The building walls are thickened and the fenestration elements set deep within them to replicate the mass of colonial adobe construction.

Mr. Glover is a longtime proponent of the courtyard plan, which he utilizes to great effect to create homes with privacy, comfort and maximum utility even on otherwise unremarkable interior parcels.

# North York Residence No.2

*Design Agency: Joseph N. Campitelli Architect Inc.*

Located just north of Downtown Toronto, ON, Canada, an existing 155㎡, one and half storey home was remodeled to extend out to a full two-storey residence. However, what started out as a small home addition and renovation soon developed into an entire building transformation, a custom home with a finished area of 270㎡.

Like most projects during the design development phase, several options and renditions were prepared to finely tune a balance between form and function. What resulted is a product which reflected the cooperation and coordination of all parties, and one which would enrich the lives of the homeowner's family.

Proposed ground floor layout included a new foyer, single car garage, side entry (corner lot exposure) and rear extension to a new larger kitchen space. At the second floor, 4 new bedrooms and 3 baths occupy the space, all accessible by a central landing/foyer designed as a single access point to each room.

For this project, JNC Architect Inc. was responsible for all aspects from planning, design, interior layout and Contract Administration during construction.

# Markham Residence Renovation No.4

*Design Agency: Joseph N. Campitelli Architect Inc.*

Looking to revitalize the home's dated identity and street appeal the homeowners retained JNC Architect Inc. to develop plans for the design and renovation work for the front façade of an existing 2-storey single-family home located in Markham, ON, Canada. Open to several options, the homeowners concluded with a decision to re-clad and wrap the entire front elevation with new contemporary materials, upgrade windows and realign a suffering roof line.

Respectively, the area for new construction was minor, at 8 ㎡ on to the existing foyer, and 10㎡ to the existing 2 car garages. However, overall the remodeling and framing work created a significant scope for the home and improvement to the front façade. A raised roof line, the construction of new gables and dormers re-oriented the front elevation and established a new grand entry to the 510㎡ home.

For this project, JNC Architect Inc. was responsible for all aspects from planning, design, interior finishes and contract administration during construction.

40

40

# Georgina Residence Addition No.1

*Design Agency: Joseph N. Campitelli Architect Inc.*

In search for the right design team to take on an expansion to their mid 19th Century cottage, the owners contacted JNC Architect Inc. to design and develop plans for a proposed room addition and interior design for their century old building. The original brick home with rubble stone foundation was planned and positioned within a golf course around the year of 1850. Fronting the main road, it overlooked a lake and backed onto a golf course with a maturing landscape. Over the past years, the golf course was eventually subdivided further into estate lots, and the original home took on a few minor additions which accommodated priorities of the time. However, it wasn't until 2008, the current owners decided to finally construct a family space specifically to take advantage and enjoy the lake front view.

A classical style of architecture was the design inspiration with the use of contemporary materials and finishes. Where once was the front door, a new area of 70㎡ was constructed in front and incorporated a roof top terrace. The main transition into the room utilized the original front entry threshold with original frame details, and brickwork surround intact. Engineered hardwood was used to line the new floor, with 2/3 painted walls and T+G softwood cladding (walls and ceiling). Douglas fir post and beams accentuate the ceiling plane while sub-dividing uses within the space. Exterior walls incorporated tall double-hung vertical sliding windows, with a 3/3 simulated true divided top panel, intended to open up the entire great room to the lake view. The Exterior cladding was completed with a low stone (local quarry) base and a board + batten pre-finished vertical wood cladding. Wood material also extended upwards to an exterior cornice molding detail and terrace railing system.

For this project, JNC Architect Inc. was responsible for all aspects from planning, design, interior finishes and contract administration during construction.

# King City Residence Addition No.2

*Design Agency: Joseph N. Campitelli Architect Inc.*

Set within a country estate location in King City, ON, Canada, the homeowners of an original 210㎡ bungalow were inspired to develop a green room to expand out into their rear yard. Once contacted, JNC Architect Inc. explored options to maximize the experience of what this space could become. Moving quickly away from conventional aluminum and glass sun room enclosures, the architect researched structural roof framing options for a cathedral ceiling, which had came to be the main inspiration of the project.

Designs were developed around the proposed plan for a one storey, 30㎡ addition with a hipped roof; a layout consistent with the existing home/roof design. Ultimately the concept concluded with a solid wood timber frame construction, for supporting posts and roof trusses, finished off in a hipped/cathedral ceiling configuration. From the main entry foyer, occupants transition from old to new through a simple opening within the existing structure. Flanked by a new stone clad wall (media center) the space opens up to an expansive view drawn between an impressive vaulted ceiling and out through the windows to a beautifully manicured rear yard landscape; accented with armored stone, pavers and plantings.

The vision concluded with the use of warm, neutral color tones and soft accent lighting for interior and exterior finishes. As a contemporary interpretation and use of traditional construction materials, the final product stands to impress as much as it comforts its owners.

For this project, JNC Architect Inc. was responsible for all aspects from planning, design, interior finishes and contract administration during construction.

# Abbots House

*Contributor: London Sotheby's International Realty*

This stunning penthouse apartment of about 6,130 s. f., has its own private indoor swimming pool with retractable roof as well as two roof terraces. This substantial penthouse has a master bedroom with a dressing room and two en-suite bathrooms, as well as two further en-suite bedrooms. There is a double height reception room, a dining room, kitchen with breakfast area, a cinema room and office. There is a lift, porterage and two underground parking spaces.

HOTEL STORIES

# Cambridge Gate

*Contributor: London Sotheby's International Realty*

This elegant penthouse apartment with magnificent views over Regent's Park is a rare find. The property offers spacious living accommodation laid out over three floors and benefits include 24 hour porterage, off street parking and a private garage for one car. There are three bedrooms, each with en-suite bathrooms with under floor heating. There is also air conditioning. The property offers two spacious roof terraces along with two balconies.

# *Nightingale House*

*Contributor: London Sotheby's International Realty*

Sotheby's International Realty presents a fine double-fronted Edwardian house situated on the renowned Harley Street, London W1. An exceptional family house steeped in history, it was designed by the great Sydney Tatchell and used to be the site of the nursing home where Florence Nightingale worked until 1854. An impressive freehold, Nightingale House is Grade II listed and built in Portland stone. It offers over 6,000 s. f. of living and entertaining space, five bedrooms, a roof terrace and garage parking for a large car. The living room features an exquisite white stone fireplace. The property also features a substantial home office. Marylebone High Street, Regents Park and Mayfair are all within close proximity.

# Park Square West

*Contributor: London Sotheby's International Realty*

A white, stucco fronted Regency house set within Regent's Park, this property has recently undergone an extensive refurbishment, and now offers a contemporary living space, while retaining its period features. There is a Master Bedroom Suite with two dressing rooms and six further bedrooms. There are three reception rooms and a library. The property has a large kitchen/dining room, divided off with stylish new work services, providing an ideal layout for entertaining.

# *Hummingbird Ranch*

*Contributor: Drew Mandile | Brooke Knapp*

Hummingbird Nest Ranch has world class facility. It covers about 123 acres, borders state park land. There is a Spanish style mansion, about 17,000 s. f. sits atop a mountain with 360-degree views of verdant riding fields and dramatic rock formations. Beamed ceilings throughout; Spanish tiles with custom inserts. Master bedroom with 13-foot ceiling and fireplace. Commercial French kitchen. Expansive French doors all opening onto verandas.

Apache Barn
Iroquois Lake
Sitting Bull Cottage

# Contributors

## Ken Tate

Recipient of the Institute of Classical Architecture's 2008 Shutze Award, and three-time winner of Southern Progress Corporation's Southern Home Award, architect Ken Tate received his Bachelor of Architecture from Auburn University in 1975. His thesis, Architecture in Search of a Soul, revealed an early interest in ancient and pre-industrial forms, pluralism, and intuition—fascinations that still inform his work. Upon completing his degree, Tate worked with architectural visionaries Bruce Goff in Texas and Sam Mockbee in Mississippi. After practicing briefly with Colonial Revival-inspired architect Richard Davis in Dallas, Tate started his own firm in 1984 in Mississippi. Since that time, he has designed more than 60 houses, including a 12,000 s.f. Federal-style compound in the Kentucky horse country, an 11,000 s.f. Norman-influenced estate in Mississippi, and a 13,000 s.f. French Colonial-inspired compound in New Orleans. Tate's work has been published in many magazines including Veranda, Southern Accents, and The Classicist, as well as books including the artist's monographs, The New Old House, and The Southern Cosmopolitan.

## Costermans Villa Projecten Belgium

COSTERMANS
VILLAPROJECTEN

At "Costermans Villa Projecten" you'll find a team of passionate people who give their all, day after day, to realize your dreams: a home with an authentic soul, a house that never stops becoming a 'home'. You've come to the right place for both classic and modern architecture, pure lines and perfect proportions are the foundation of every design. The art of combinations in balance, without overdoing it are being exercised on daily basis.

For a demanding clientele the design process of every home starts out with a blank page, to result in a flawless delivery with a satisfied customer as the most important goal. Every customer is unique, everyone has his specific requests and living habits who need to be translated, obtaining an organization within a pleasant environment with the right atmosphere. Designing and developing contemporary and modern homes with characteristic accents such as historic building materials and exclusive custom work, result in a unique experience of the object.

## James Glover

Mr. Glover's passion for Architecture has deep roots. Being a son of respected builders in Southern California, he completed his first project, as primary superintendent, at sixteen years old.

Cultivating a fertile base of knowledge in the arts, particularly the histories of Architecture and Construction, he has developed his own, singular mode of expression embodying centuries of accumulated knowledge. For him, Architecture is more than technology. It is Craft.

"Every client's vision is unique", says Mr. Glover. "So I approach each project with an open mind. My goal is to create environments that balance the occupant's lifestyle with nature by carefully blending the internal and external spaces… To create environments that are light, airy, maximize views, provide an indoor-outdoor living experience and harmonize with the site."

## Joseph N. Campitelli Architect Inc.

Joseph N. Campitelli Architect Inc. is a design firm based in Markham, Ontario, Canada, offering a full range of Architectural services; from technical drafting, computer rendering, innovative designs in building and planning, to project management and effective construction options. With several years of experience, Joseph Campitelli has in the past developed several projects varying in scale from custom homes, community planning and high-rise condominiums, commercial and retail developments.

The office was established in 2004 by Joseph Campitelli. An award winning and creative architect, who has combined his years of experience with an ambition to provide personal service and to strive for architectural excellence. Unified but diverse talents and years of broad training has strengthened the firm's role and responsibility to current environmental issues, striving to exceed current standards within the industry. Committed to being resourceful and in the completion of a vision, the firm provides the best in service and innovation.

In partnership with colleagues in Dubai, UAE, services are offered internationally with comprehensive design solutions in architecture, interior design, and planning for all project types and scales. JNC Architect and its associates look to provide services locally as well as expand internationally, where their extensive resources serve clients and where there is a potential to grow.

## London Sotheby's International Realty

Mayfair office

Mayfair, Belgravia and the surrounding borough of Westminster are iconic districts of central London. The areas are often seen as the cultural centre of the city where the national museums, galleries, theatres, concert halls and heritage buildings are found. The majority of embassies, London's most famous hotels, Michelin starred restaurants and the internationally renowned shopping of Bond Street are also all located here. With a variety of architectural styles; from stately 19th century mansion houses in Mayfair to the elegant "piano nobile" apartments on the leafy garden squares of Belgravia, the borough of Westminster has a feeling of old world prestige. The residential property here is generally considered to be some of the finest in the world. The Mayfair office in London is adjacent to Sotheby's auction house with which it maintains a strong relationship. Centrally located just off Bond Street, the office provides bespoke services to a global clientele who want to sell or buy in Mayfair, Belgravia, St James', Marylebone, Regent's Park and increasingly in the more family orientated areas of St John's Wood and Hampstead.

## Drew Mandile | Brooke Knapp

Drew Mandile

With over 35 years of real estate experience exclusive to the Westside of Los Angeles, Drew Mandile has sold or been involved in more than 2,000 residential transactions. As a young man, Drew began in Beverly Hills in a boutique agency. Drew's sales included being involved in the 24-home Baywood Estates Development in Glenridge. He then opened his own real estate firm and design center on Camden Drive with 35 associate agents. He later sold the firm to a large regional company.

Brooke Knapp

Having joined Sotheby's International Realty in 1996, Brooke has had many notable real estate achievements with partner Drew Mandile, including the sale of Moraga Canyon, over 260 acres of raw land in Bel Air, and the self-sale of Owlwood and the adjacent Jayne Mansfield Estate, totaling 10 acres. This $35 million transaction was a record sale. She is master certified with GRI designations.

Drew Mandile | Brooke Knapp

Since 1996, Drew Mandile and Brooke Knapp have been one of the most successful real-estate teams in Beverly Hills, working for Sotheby's International Realty on Wilshire Blvd.

We are committed to quality in our business practices. We hold ourselves to the highest ethical standards. We tell the truth to our clients, our business partners, and each other.

The partnership of Drew Mandile & Brooke Knapp is founded upon the trust and confidence of our clients. Our mission is to help people improve their lives through real estate.

## Acknowledgements

We would like to thank all the designers and companies who made a significant contribution to the compilation of this book. Without them, this project would not have been possible. We would also like to thank many other people whose names do not appear in the credits, but who provided valuable input and support for the project from the start to its completion.

## Future Editions

If you would like to contribute to future Artpower publications, please email your details to: info@artpower.com.cn.